Reading Essentials®
in Science

COMMUNITIES OF LIFE

Mountains

JANE HURWITZ

PERFECTION LEARNING®

Editorial Director:	Susan C. Thies
Editor:	Mary L. Bush
Design Director:	Randy Messer
Book Design:	Brianne Osborn, Emily J. Greazel
Cover Design:	Michael A. Aspengren

A special thanks to the following for their scientific review of the book:
Paul Pistek, Instructor of Biological Sciences, North Iowa Area Community College
Jeffrey Bush, Field Engineer, Vessco, Inc.

To Olivia

Image Credits:
© George D. Lepp/CORBIS: p. 17 (top); © O. Alamany & E. Vicens/CORBIS: p. 19; © David Keaton/CORBIS: p. 25; © Associated Press/The Salt Lake Tribune: p. 22

Corel Professional Photos: front cover (background, bottom left, bottom center), pp. 3, 4–5, 6, 7, 8–9, 14–15, 18, 23, 24, 26–27, 31, 32–33, 37; Digital Stock: pp. 20–21, 38–39; MapArt: p. 21; NOAA: pp. 12, 29; Perfection Learning Corporation: pp. 9, 10; Photos.com: front cover (bottom right), back cover, pp. 11, 15, 16, 17 (bottom), 28, 30, 32, 35, 40

For information, contact
Perfection Learning® Corporation
1000 North Second Avenue, P.O. Box 500
Logan, Iowa 51546-0500.
Phone: 1-800-831-4190
Fax: 1-800-543-2745
perfectionlearning.com

2 3 4 5 6 7 PP 09 08 07 06 05 04

ISBN 0-7891-6099-4

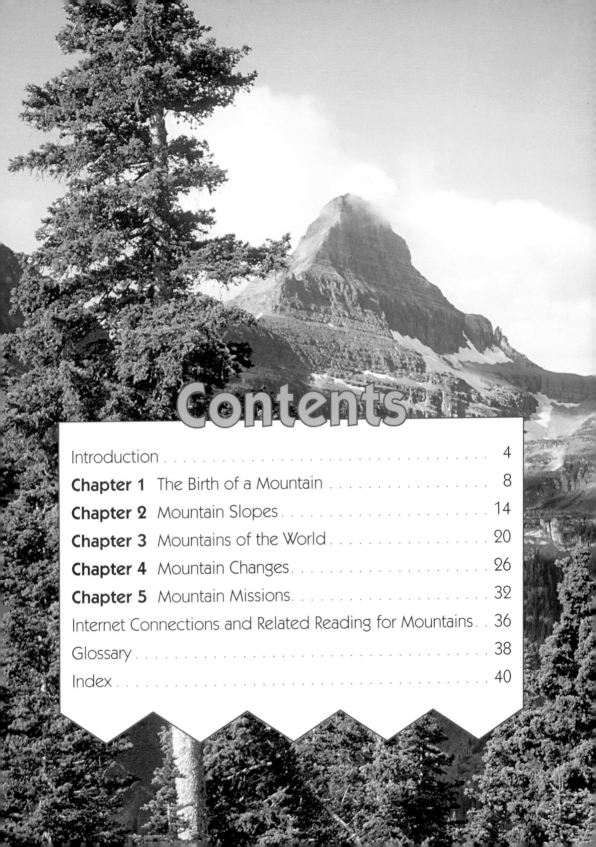

Contents

Introduction

If someone asked you to describe the area where you live, what would you say? Do you live in a desert region where it's hot and dry? a forest area with lots of evergreen trees? near a hot, wet tropical rain forest? How would you describe the temperature, sunlight, and rainfall in your hometown? What plants and animals live there?

Biomes

What you are describing is a **biome**. A biome is an **environment** with unique features. For example, an ocean biome has salt water. A **tundra** biome is cold and dry, and often the ground is frozen year-round.

There are many types of biomes, including desert, mountain, tundra, forest, grassland, ocean (saltwater), freshwater, and rain forest. Ecologists have noticed that the same biomes can appear in very different places. Deserts, for example, are found in both hot and cold locations. But even though they are in different parts of the world, all deserts share some characteristics.

Each biome has its own special plant life. Think about the different plants found in a desert, a rain forest, and a grassland. Cactuses grow in the desert. Palm trees grow in the rain forest. A variety of grasses cover the grassland.

Biomes are also identified by how plants and animals must **adapt** in order to live there. To live in an ocean biome, plants and animals must be able to live in salt water. In a desert, the wildlife must be able to survive long periods without water. Each biome has its own unique environment to which the plants and animals must adapt.

Ecosystems

Ecologists have also determined that certain groups of plants and animals tend to live

Baboon

together. These groups of living creatures interact with the nonliving parts of the environment, such as rocks or sand. Groups of living creatures that interact with one another and their surroundings are called **ecosystems**.

Each biome is made up of many ecosystems. In an ocean, there are different ecosystems living in **coral reefs**, cold Arctic waters, and deep underwater **trenches**. Each layer of a tropical rain forest has its own ecosystems.

Working together, the ecosystems form **communities** of life within each of the biomes.

Rocky Mountain bighorn sheep live at altitudes of 5000 to 10,000 feet.

Chapter 1

The Birth of a Mountain

Has anyone ever said to you "don't make a mountain out of a molehill"? If so, someone was telling you not to make a big problem out of a small one. But what is the real difference between a mountain and a hill? A hill is defined as "an area of raised land smaller than a mountain." A mountain is defined as "an area of raised land larger than a hill." That really doesn't help much, does it? While what is "bigger" and "smaller" may differ to people, it is generally agreed upon that anything shorter than 1000 feet is a hill and anything taller than 1000 feet is a mountain.

In the Beginning

All mountains are born when rocks beneath the Earth's surface move. The rocks that make up the Earth form in layers, like the layers of skin on an onion. The inner layer is hot rock called the *core*. The middle layer is melted rock called the *mantle*. The thin, hard outer layer of the Earth is called the *crust*. The crust is thinnest under the oceans and thickest under the continents.

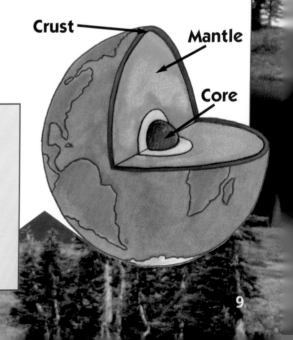

Crust — Mantle — Core

How to Figure Out a Mountain's Height

The height of a mountain usually refers to how many feet above sea level the mountain stands. So a mountain that is 10,000 feet rises 10,000 feet above the sea. The measurement doesn't include any part of the mountain that might be underwater.

Even though the rocks in the Earth's crust are very hard, they are not one solid piece. The rocky crust is broken into pieces called *plates*. The Earth's crust is divided into eight large plates and a few smaller ones. These plates move because they float on a layer of liquid rock called *magma*.

The United States is on the North American Plate. Europe is located on the Eurasian Plate. The Appalachian Mountains in the United States were formed when the North American Plate collided with the Eurasian Plate, pushing land upward.

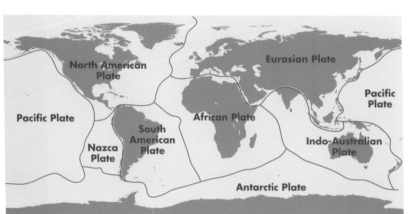

Under the ocean, there are cracks between the plates. Magma oozes through these cracks, causing the plates to move farther apart. When the magma hardens, new land is created. Over millions of years, the movement of the plates results in large changes to the surface of the Earth. Mountains are created as the plates of the Earth's crust push together or pull apart.

It takes thousands of years for the movement of the plates to create mountains. Depending on the movement of the plates, different types of mountains are created. Some of the more common types are fold mountains, fault mountains, dome mountains, and volcanic mountains.

Fold Mountains

When two plates in the Earth's crust bump into each

other, pressure is created. It takes thousands of years, but eventually the plates push so hard against each other that they bend. Rocks in the plates are folded by the pressure and pushed upward, creating fold mountains.

The largest mountain **ranges** in the world are fold mountains. The Himalayas, the Rockies, and the Andes are all examples of fold mountains.

Fault Mountains

When two plates in the Earth's crust pull apart from each other, pressure is released. Layers of rock in the plates break. A crack called a *fault* forms in the Earth's crust. Large blocks of rock rise up and slip down along the fault line, forming mountains. Fault mountains often have a gentle **slope** on one side. The other side is usually a very steep wall.

When the North American Plate moved away from the Pacific Plate, the San Andreas Fault formed along the coast of California. The Sierra Nevada are a result of this fault.

Dome Mountains

Sometimes magma from deep within the Earth pushes against the Earth's crust. The magma oozes upward but doesn't break through the surface. As the magma hardens into rock, it pushes upward. The land above becomes a dome-shaped mountain.

The Black Hills of South Dakota formed 60 to 70 million years ago when a fountain of magma pushed the land upward. These mountains contain many natural features, such as caves, cliffs, canyons, waterfalls, and the remains of ancient volcanoes.

Sierra Nevada

13

Volcanic Mountains

When magma manages to find a crack in the Earth's surface, it explodes out of the ground as **lava**. As the lava hardens around the opening, a mountain is born. The final shape of the volcanic mountain depends on many things—how much magma escapes, how fast the lava flows, and how often and how long eruptions occur. Each time a volcano erupts, the shape of the mountain changes.

Mount Fuji is a volcanic mountain that is currently **inactive**. Rising 12,389 feet, Mount Fuji is the highest mountain in Japan. Nearly 200,000 people climb this mountain every year.

Volcanic mountains that form under the sea are often hidden from view. At other times, they grow to such a height that they rise above the water and become islands.

The longest and tallest mountain chain on Earth is made up of volcanic mountains. Known as the Mid-Atlantic Ridge, this chain of volcanoes stretches for 12,000 miles. **Peaks** of some of the volcanoes reach 36,000 feet high, which is about 24 times the height of the Empire State Building. Most of these giant mountains are completely covered by water. However, a few rise above the surface, forming islands. The Mid-Atlantic Ridge is located in the middle of the Atlantic Ocean between North America and Europe. This is where two of the Earth's plates are pulling away from each other.

Spirits on the Mountain

Mount Fuji is a sacred spot for people of the Buddhist and Shinto religions. Buddhists and Shintos believe that spirits live on Mount Fuji. It is also believed that as people climb higher toward the sky, their religious beliefs grow stronger.

The Hawaiian Islands are also volcanoes that formed on the bottom of the ocean. These mountains rise 30,000 feet above the Earth's crust, creating the eight islands of Hawaii.

As They Grow

A mountain may grow to just one large peak or become part of a group, or range, of mountains. The Himalayas are a range of tall mountains in Asia. As with all very tall mountains above sea level, the tops, or peaks, of the Himalayas are covered with snow all year long. The treeless, rocky peaks of the Himalayas reach an average of 20,000 feet.

The Rocky Mountains range in North America is 2980 miles long. The Andes Mountains range in South America is 4475 miles long. When mountain ranges join together, they become a **chain**. The Rockies and the Andes form a mountain chain known as the Cordilleras.

Whether a chain, a range, or a single peak, many mountains are worn down by time. Weather and water wear away mountains until their peaks become lower and more rounded. The Appalachian Mountains in the eastern United States form one of the oldest mountain ranges on Earth. These weathered mountains are millions of years old. Estimated to be nearly 20,000 feet at one time, the highest peaks now stand at approximately 6000 to 7000 feet. These old mountains are covered with forests and grasslands that have grown thick over time.

Tour of the Tallest

By visiting Alaska, California, Colorado, and Washington, you can see the 50 highest mountains in the United States. These mountains include Mount Rainier, Mount McKinley, and peaks in the Rocky Mountains and the Sierra Nevada.

Chapter 2

Mountain Slopes

As you climb a mountain slope, you'll notice different ecosystems at various heights. The mountain biome is actually a combination of several other biomes. Climate differences at each level of a mountain slope create perfect places for grassland, savanna, forest, rain forest, and tundra biomes to thrive. Rivers that run from the tops of mountains to their bottoms are often surrounded by freshwater biomes. Only on a mountain can you find such a variety of communities.

Lower Slopes

The lower slopes of mountains are the warmest zones. Trees, grasses, plants, and flowers thrive here. Rivers run freely, providing freshwater for plants and animals. Biomes such as savannas, grasslands, and rain forests can be found at the base of many mountains.

Mount Kilimanjaro is the highest mountain on the African continent. At the base of this mountain is a savanna. A savanna is an area of grassy land with a few trees. On the African savanna, the weather is warm most of the year. Giraffes, lions, and zebras roam the grassy plains.

Farmers on small farms called *shambas* grow bananas and vegetables on the savanna. The flat land is also used for coffee plantations. Farmers try to raise cattle along the lower slopes as well but struggle due to lack of rain.

Above the savanna, the slopes of Kilimanjaro rise into a rain forest. Giant ferns grow among fragrant eucalyptus trees. Baboons and colobus monkeys play in the trees. Antelopes, bushpigs, and elephants walk along the rain forest floor.

The warm climate at the base of Africa's highest mountain is quite different from the lower slopes of Europe's highest peak, Mount Elbrus in Russia. Since Mount Elbrus is far away from the equator, the lower slopes are never as warm as those of Kilimanjaro's. Trees such as birch, aspen, and ash grow along the base of Mount Elbrus. While trees in the rain forest keep their green leaves all year long, the trees in the **deciduous** forests of Mount Elbrus react to the cooler climate. In the fall, they turn brilliant shades of red and yellow.

Then they shed their leaves in the winter.

Himalayan black bears live in the forests on the lower slopes of the Himalayas. These bears are good tree climbers and eat many of the plants found in the forest. Himalayan black bears eat as much as possible in the fall to store up fat for the cold winters. Then they spend the winter sleeping in caves or holes in trees.

Can You Guess Where They Live?

Many animals that live on mountain slopes are named for their home. These include the mountain lion, mountain goat, mountain bluebird, and mountain squirrel.

Mountain goats live on the steep mountain ranges of northwestern North America.

Mountain aven

Middle Slopes

Along the middle slopes of most mountains, the climate is cooler. **Conifer** trees, such as pine, spruce, larch, and fir, grow on these areas of the mountain. Most of these trees have needlelike leaves to help them survive the winters. The needles store water, which allows the leaves to continue making food in cold temperatures. Farther up the middle slopes, the trees grow smaller due to the lower levels of oxygen and even cooler temperatures.

Even though it's cold and the air is thin, people and animals can live along the middle mountain slopes. The Alps form a huge mountain range in Switzerland. Farmers have cleared part of the coniferous, or evergreen, forest found on the middle slopes to create hay fields. In the summer, cows and goats graze higher up the mountains while grass is grown along the middle slopes. In the fall, the grass is harvested as hay. The animals eat the hay while they pass the cold, snowy winter in the shelter of a barn.

17

their homes in the Alps. These birds have a crest of black-tipped feathers on their heads and long bills that they use to protect themselves and their nests.

Upper Slopes

Higher up the mountain, the slopes are much colder. The oxygen in the air decreases further. The climate becomes very cold, windy, and dry. Trees and shrubs can no longer survive. These conditions are ideal for the tundra biome. The tundra is a cold, treeless area with little moisture.

Mountain Houses

Houses in the Alps are called *chalets*. They have high, sloping roofs that hang down far over the tops of the houses' walls. These special roofs help prevent snow from falling on the windows and doors. Chalet windows are usually small, which helps decrease heat loss.

The End of the Line

The tree line, or timberline, is the area on a mountain above which no trees grow.

The forests of the middle slopes provide homes for many creatures. In the Alps, chamois leap among the trees, rocks, and cliffs. These animals have spongy pads under their hooves to help them grip rocks. Alpine ibex are mountain goats with large horns. During the summer, ibex have short, fuzzy coats. In cooler temperatures, their coats grow thicker to keep them warm. Birds called *hoopoes* also make

This upper area of the mountain is often called the *alpine zone*. **Organisms**, such as mosses and **lichens**, struggle to grow in the harsh climate. Small creatures, such as spiders, beetles, and rodents, live in this area. Birds of prey, such as eagles, vultures, and buzzards, hunt in the alpine zone.

In the Andes Mountains of South America, several unique animals can survive in the tundra environment. The vicuna is a small, woolly animal that eats the tiny plants that can grow in the cold temperatures. Vicunas can breathe in the thin mountain air because their blood holds three times more oxygen than human blood does. Alpacas are related to vicunas. The long, thick hair of the alpacas keeps them warm on the upper mountain slopes. People actually raise alpacas for their hair, which can be made into a soft, warm wool.

Above the alpine zone is the arctic zone. The air at this level contains less than half of the oxygen found at sea level, making it difficult for animals and humans to breathe. In the arctic zone, glaciers form into permanent ice fields that can be as great as 30 feet high.

The Karakoram Mountains of Pakistan have many of the world's highest peaks and largest glaciers. The sharp, craggy peaks contain the most glacial ice found anywhere in the world outside the North and South Poles. Water melting from these glaciers flows downhill to the Indus River. Without the Indus River, the people of Pakistan could not farm the dry plains that surround the mountains.

Even though people don't live in the upper slopes of the Karakoram, people do visit there. The second-highest mountain peak in the world, called K2, is in the Karakoram Mountains. This peak challenges many experienced mountain climbers.

Alpine ibex live and travel in herds of 5 to 20 members in the craggy terrain of the Alps.

Chapter 3

Mountains of the World

Mountains can be found on all of the seven continents. The highest mountain peaks, or **summits**, on each continent have been placed in a special group. They are called the *seven summits*.

Over the years, however, there have been arguments about the seventh summit. Those who consider Australia a continent believe that Mount Kosciuszko is the seventh summit. Others believe that the island of Australia is part of a continent called Australasia or Oceania, which includes Australia and the group of islands surrounding it. Those people believe that Indonesia's Carstensz Pyramid is the seventh summit.

Aconcagua

Mount Aconcagua is the highest peak in South America. Rising more than 22,000 feet, this summit is an **extinct** volcano found in Argentina. It is part of the Andes Mountains range. The top of Mount Aconcagua is covered with ice and snow year-round.

Carstensz Pyramid

Carstensz Pyramid is located in the Jayawijaya Mountains range on the island of Indonesia. This limestone summit rises more than 16,000 feet. Because it is close to the equator, the weather on the mountain stays the same

Dick Bass after climbing Mount Everest

most of the year. Rain falls almost daily. Snow and ice top the peak. Carstensz Pyramid is named for Jan Carstensz, the first explorer to spot the peak.

Elbrus

The Caucasus Mountains of Russia are home to Mount Elbrus. Rising more than 18,000 feet on its western face, Mount Elbrus is the highest mountain on the European continent. The mountain is actually the two-headed cone, or top, of an inactive volcano. It has east and west summits, which differ in height by about 100 feet. Mount Elbrus is made of hard rock and magma formations. The climate is often unpredictable, with bursts of sudden winds, snow, and ice.

Everest

The Himalayas pass through the countries of Tibet and Nepal in Asia. Mount Everest is the tallest peak in the Himalayas. At more than 29,000 feet, it is also the world's tallest summit. Avalanches, steep slopes, wind, and ice make Mount Everest a difficult mountain to climb.

Mount Everest is called Chomolangma by the Tibetan people. Chomolangma means "Goddess Mother of the Snows." The Nepalese people call this summit Sagarmatha, or "Mother of the Universe."

Himalayan Homes

In many parts of the Himalayas, temples and homes for worshipers have been built. Teachers and students of the Buddhist religion live in groups of houses called *monasteries*. These high mountain homes provide a place for quiet study far away from modern life.

The Himalayas

Kilimanjaro

In Tanzania, Africa, Mount Kilimanjaro rises from the dry plains to a height of more than 19,000 feet. The base of Kilimanjaro is a perfect location to take an African safari. Lions, tigers, giraffes, elephants, and cheetahs roam the hot, grassy land.

Kilimanjaro is an extinct volcano with two peaks. One peak, Kibo, is the highest point on the continent. It is always covered with snow, ice, and glaciers. The other peak, Mawnzi, is almost 2000 feet shorter and does not have glaciers.

Kosciuszko

Mount Kosciuszko is Australia's highest summit. It is in the Australian Alps. This peak towers over Australia at more than 7000 feet.

In the summer, the top of the grassy, dome-shaped Kosciuszko can be reached in less than a day. Snow in the winter makes the climb more difficult. The mountain pygmy opossum and the corroboree frog are two rare animals found on this mountain's slopes.

An elephant near Mount Kilimanjaro

McKinley

In most parts of the world, people depend on the freshwater that flows from mountains. Water melting from the glaciers on Mount McKinley, the highest peak in North America, flows into an area used by two-thirds of Alaska's population. The mountain stands more than 20,000 feet tall.

Mount McKinley is located in Alaska's Denali Wilderness. Denali is the mountain's Native American name, which means "The Great One" or "The High One." The summit was named Mount McKinley in honor of William McKinley, America's twenty-fifth president.

Vinson Massif

Vinson Massif is the highest peak in Antarctica. It rises into a pyramid shape more than 16,000 feet above an icy, frozen land. In fact, the base of this high mountain sits on a giant shelf of ice. Vinson Massif is located 600 miles from the South Pole. The ice, snow, and freezing temperatures make this the most difficult of the seven summits to explore.

Difficult Dreams

Many people dream of climbing all of the seven summits, but only about 400 people have ever made it to the top of Vinson Massif.

Chapter 4

Mountain Changes

Mountains can take millions of years to form. During that time, the mountains are always changing. But even after their formation, mountains continue to change. Heavy snowfalls, pounding rains, and changes in temperature affect the shape of mountains every day. Changes are also caused by humans.

Nature's Changes
Weathering

Most mountains are under attack from the forces of nature. Water, temperature, and wind work away at the surfaces of these mountains.

Chemicals in the air and water also help to break down a mountain's rocky surface. These forces, called *weathering*, cause mountains to crack and crumble.

Water and changing temperatures can be a destructive combination for mountain slopes. Water from glaciers, streams, and rivers runs into cracks in rocks on the mountain slopes. When temperatures drop, the water freezes and expands. This can cause the cracks to grow even bigger or the rocks to split into pieces.

The Grand Canyon in Arizona was created by many years of water erosion.

As weathering occurs on mountain slopes, pieces of rock and soil are carried away. This is called *erosion*. Water, wind, ice, and snow move these pieces of rock and soil down the mountain. Water carries some of the material into rivers. Wind can blow lighter particles off the mountain slopes to the surrounding area.

Landslides

Piles of weathered rocks and soil can become very heavy. Eventually, gravity pulls them down the mountain slope. Large amounts of weathered materials moving down a mountain can cause a landslide, also known as a *mudslide*.

The mountains on the island of Java in the Pacific Ocean experience landslides from heavy rains. Mud and boulders mixed with rushing water flow down the mountainside. Roads, villages, people, and forests in the mudslide's path are buried. These landslides are worse when tropical forests on the mountain slopes are cleared for farming or timber. Without trees to stop the flow of the mud and rocks, the lower mountains are easily washed away in the rushing force of the mudslide.

Volcanoes

The mountains on Java also see changes because of the many active volcanoes there. Treeless on top, many of the blue-gray mountains still explode. Much like a mudslide, the lava from an erupting volcano changes or destroys everything in its path. Yet unlike a mudslide, the lava also creates new, productive land.

New plant growth on a cold lava flow in Hawaii

Lava is rich in **nutrients**. After an eruption, rainwater washes weathered materials like ash and soil over the lava. Over time, the lava breaks down and mixes with the weathered materials to form a rich soil. In the end, the lava creates mountain slopes that support thick forest growth or fertile farmland. Volcanoes on Java have made the lower mountain slopes perfect for rice farming.

Avalanches

Colder mountain areas may experience snowslides called *avalanches*. Avalanches occur when layers of snow on steep slopes are pulled downhill by gravity, wind, or an object such as a mountain climber, skier, or animal. As the snow moves down the mountain, it picks up bits of ice, rocks, soil, and other materials. Anything in the avalanche's path is buried beneath the snow and **debris**.

When the avalanche is over, life on the mountain has changed. Weathered materials carried by the tumbling snow have been left along the path or at the mountain's base. Trees have been damaged. Animal habitats have been destroyed. Plants, animals, and even humans may have been killed.

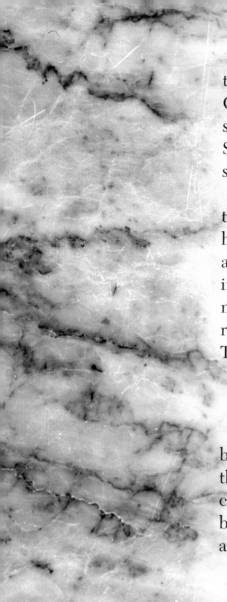

The Caucasus Mountains in Russia make up a tall mountain range that divides Europe and Asia. Grasslands and wildflowers grow on the lower slopes. Farther up, however, the weather is harsh. Snow and high winds are possible even in the summer. Avalanches occur with little warning.

A loud roar that sounds like thunder is often the only warning people have before an avalanche hits. The roar is caused by chunks of ice as large as trucks sliding down the Caucasus range. Trees in the avalanche's path snap in half like matchsticks. When the avalanche is over, tons of rock, mud, ice, and snow have shifted downslope. The face of the mountain is greatly changed.

Human-Made Changes

Some of the changes on mountains are caused by humans. Mining on mountain slopes damages the rock layers. Cutting, digging, and dynamiting changes the landscape forever. Clearing land for buildings or recreational areas destroys plants, animals, and habitats.

When stone is cut from a mountain, a sharp gap is left in the mountainside. In the Alps of northern Italy, shiny white gaps are visible where marble has been cut from the mountains. Near the Italian town of Carrara, marble has been carefully cut from the mountainside for hundreds of years. It is shipped all over the world. Marble is also cut from the mountains in China, Afghanistan, and Malaysia.

Marble Mining

Marble has been mined from the Alps for more than 2000 years. The famous sculptor, Michelangelo, used marble from Carrara more than 500 years ago.

Mountains have also been used to create works of art. Mount Rushmore and Stone Mountain are two examples of mountains that have been sculpted by human hands. The two mountains have many things in common. Both are dome mountains formed by magma. Both mountains have a layer of **granite** on their surfaces. Both mountains hold carvings that represent America's past. In fact, both mountains were changed by the same man.

Gutzon Borglum designed and sculpted the image of three Southern Civil War heroes into the granite surface of Stone Mountain. Located near Atlanta, Georgia, the carved surface is larger than a football field. Originally started in 1915, Borglum left the project in 1925, leaving the side of the mountain scarred and unfinished. Years later, the monument was finally finished.

After Stone Mountain, Borglum moved on to bigger projects. In 1927, he began working on Mount Rushmore. Borglum sculpted the heads of four American presidents— George Washington, Thomas Jefferson, Theodore Roosevelt, and Abraham Lincoln. The giant presidential heads are carved into the granite surface of Mount Rushmore in South Dakota. The sculpture was made to honor the first 150 years of America's history.

Today, both monuments are surrounded by carefully maintained mountain land. Millions of people visit the monuments each year to enjoy the beauty created by the combination of man and mountain.

Each president on Mount Rushmore would stand about 465 feet tall if his entire body were shown.

Chapter 5

Mountain Missions

One-fifth of the Earth's land is covered by mountains. One-tenth of the world's people live on mountains. Billions of people depend on mountain ecosystems for food, water, electricity, wood, and **minerals**. More than half of the freshwater in the world begins in the mountains as snow, ice, or rain. Clearly, most of the world is affected in one way or another by the mountain biome.

As the number of people in the world increases, the demands placed on mountain ecosystems grow too. Making use of mountain resources while at the same time protecting this important biome is an enormous mission. Several issues threaten the success of this challenge.

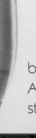

A Drink from the Mountains

Some of the largest rivers in the world begin at high mountain elevations. The Amazon, Nile, Indus, and Missouri Rivers all start out as high mountain lakes or glaciers.

Deforestation

Deforestation is the removal of trees from an area, including mountain slopes, for timber, construction, farmland, and other uses. Without trees, the forest floor is left bare, without protection from wind and rain. Soil is easily eroded, or carried away, taking with it valuable nutrients that plants depend on. When plant life decreases or disappears, many mountain creatures lose an important source of food.

Soil erosion doesn't only destroy mountain ecosystems. It also affects lands beyond the mountainside. Soil carried off mountains often clogs up rivers and streams. This can create flooding in areas beneath the mountain.

Deforestation can also speed up nature's changes. Without trees on a mountain slope, the barriers to mudslides or avalanches are removed. There is nothing left to stop the flow of rocks, soil, snow, and ice as it rushes down the mountain. On the volcanic mountains in Java, for example, mudslides are worsened when forests along the lower slopes are harvested for wood.

Art or Gift of Nature?

Stone Mountain, Mount Rushmore, and other mountain sculptures like them are considered great works of art by many. But others believe that creating art out of mountainsides destroys the natural beauty and purpose of the mountains.

Concerned people fear the effects of art on the mountain environment. When changes are made on mountain slopes, the plants and animals that live there are endangered. Clearing areas of a slope for sculptures destroys habitats. Changing the natural land features can create problems as well. On Mount Rushmore, lichens cling to the curves of Abraham Lincoln's sculpted granite beard. Over time, these organisms cause the rock to crack and crumble. This erosion can damage plant and animal life along the mountain slopes.

Some groups of people believe that mountains are sacred and should not be altered by human hands. The Lakota Sioux are Native Americans who lived in the Black Hills of South Dakota long before the Europeans arrived. To the Lakota, Mount Rushmore is a sacred place, not one that should have been used to create a work of art.

With opinions on both sides, no easy solutions have been found. Questions still remain. Who should decide if a mountain should be preserved in its natural state or be used to create a work of art? Who will take responsibility for the destruction to the mountain biome? Which is more important—art or nature?

Visitors Destroy the View

More people visit mountain areas for recreation each year. Skiers, snowboarders, hikers, and climbers flock to the mountains for fun. But with these visitors come problems. As tourists ski, board, or climb up and down the mountains, they leave behind litter, damaged habitats, and further erosion.

Mount Kenya, the second highest peak in Kenya, Africa, has a huge litter problem created by hikers and climbers. One creative solution that is being considered is to make each person going up the mountain buy a trash bag from a park ranger. If the bag is returned full of trash at the end of the hike, then the hiker's money will be returned.

There are also questions about making changes to mountains to make recreational activities more comfortable and convenient. For example, the number of visitors climbing Mount Fuji increases every year. People have begun to wonder if public facilities, such as toilets and restaurants, should be built on the mountain. Some argue that this would change ecosystems and damage habitats. It would also destroy the sacredness of the mountains for groups such as the Buddhists and Shintos.

The towering peaks of mountains provide beauty and resources for people, plants, and animals. It is important to find a balance between using the mountains for resources, art, or recreation and preserving the mountain environment. Making sure that the world's mountains don't become molehills is a valuable mission for everyone.

Internet Connections and Related Reading for Mountains

http://www.globaleye.org.uk/primary/focuson/intro.html
Find out about the world's highest mountains with this file of mountain facts. Explore Machu Picchu and Mount Everest. Then take a quiz to see if you know the locations of the world's highest mountain ranges.

http://www.4learning.co.uk/essentials/geography/units/mount_bi.shtml
Learn the "essentials of the mountain environment" at this site. Includes information, activities, and a quiz.

http://www.panda.org/news_facts/education/virtual_wildlife/wild_places/mountains.cfm
Examine the mountain biome—its formation, ecosystems, and plant and animal life.

http://www.hautes-alpes.net/site/uk-ete/connais.htm
Find out what you know about mountains at this fun site. Check out the information on mountain zones, plants, and animals. Puzzles and games add to your knowledge.

http://www.allaboutnature.com/biomes/
This site provides an overview of the many different biomes found on mountain slopes (grassland, savanna, rain forest, coniferous forest, deciduous forest, tundra). Click on any of these biomes for information on the biome and the animals that live there.

Endangered Mountain Animals by Dave Taylor. Poachers and changing population patterns are leaving less room for a multitude of rare and wonderful wildlife species that inhabit the mountain regions of the world. Crabtree Publishing, 1992. [RL 4 IL 3–7] (4391401 PB 4391402 CC)

How Mountains Are Made by Kathleen Weidner Zoehfeld. Describes plate tectonic theory and how the forces of nature shape our world. HarperCollins, 1995. [RL 2 IL K–4] (4799601 PB 4799602 CC)

Mountains by Neil Morris. Looks at the natural characteristics and ecology of mountains. Includes index and glossary. Crabtree Publishing, 1996. [RL 4 IL 2–5] (4974401 PB 4974402 CC)

Mountains by Seymour Simon. From the snowy Alps to the Appalachians' worn peaks, mountains are a dramatic reminder of ages past and ages to come. William Morrow, 1997. [RL 3 IL K–4] (5543601 PB)

What Is a Biome? by Bobbie Kalman. This book introduces biomes, showing and describing the main kinds and discussing their location, climate, and plant and animal life. Crabtree Publishing, 1998. [RL 3 IL 2–5] (5729401 PB 5729402 CC)

• RL = Reading Level
• IL = Interest Level
Perfection Learning's catalog numbers are included for your ordering convenience.
PB indicates paperback. CC indicates Cover Craft.

Glossary

active (AK tiv) having erupted recently; likely to erupt again in the future

adapt (uh DAPT) to learn to successfully live in an environment (see separate entry for *environment*)

biome (BEYE ohm) environment with unique features (see separate entry for *environment*)

chain (chayn) two or more groups of mountains linked together

community (kuh MYOU nuh tee) organisms that live together in a particular location (see separate entry for *organism*)

conifer (CON uh fer) type of tree that produces cones; most conifers have needlelike leaves that remain green all year long

coral reef (KOR uhl reef) rocky area in warm, shallow ocean waters created from the remains of animals called *polyps*

debris (duh BREE) remains of something broken down or destroyed

deciduous (dee SIJ oo uhs) type of tree that cycles through stages, losing its leaves in the fall

ecosystem (EE koh sis tuhm) group of living creatures that interact with one another and their surroundings

environment (en VEYE er muhnt) set of conditions found in a certain area; surroundings

extinct (ek STINGKT) not having erupted for a long time; not likely to erupt again

granite (GRAN it) hard rock with visible crystals

inactive (in AK tiv) not having erupted for a long time

lava (LAH vah) melted rock that rises above the Earth's surface

lichen (LEYE kuhn) organism that grows on another surface, such as a rock (see separate entry for *organism*)

mineral (MIN er uhl) nonliving material found in nature

nutrient (NOO tree ent) material that living things need to live and grow

organism (OR guh niz uhm) living thing

peak (peek) top of a mountain

range (raynj) group of mountains

slope (slohp) upward or downward slant or side of a mountain

summit (SUH mit) highest peak on a mountain (see separate entry for *peak*)

trench (trench) deep canyon, or valley, on the ocean floor

tundra (TUHN druh) treeless region with soil that is often frozen year-round

Index

TRADITIONAL TALES

from

NORTH AMERICA

Vic Parker

Based on myths and legends retold by

Philip Ardagh

Illustrated by

Olivia Rayner

Thameside Press

U.S. publication copyright © 2001 Thameside Press.

International copyright reserved in all countries.
No part of this book may be reproduced in any
form without written permission from the publisher.

Distributed in the United States by
Smart Apple Media
1980 Lookout Drive
North Mankato, MN 56003

Editor: Stephanie Turnbull
Designer: Zoë Quayle
Educational consultant: Margaret Bellwood

Library of Congress Cataloging-in-Publication Data

Parker, Vic.
    North America / written by Vic Parker.
        p. cm. -- (Traditional tales from around the world)
    Summary: A collection of myths and legends from Native Americans,
European settlers, and African slaves.
    ISBN 1-930643-39-X
    1. Tales--United States. [1. Folklore--United States.] I. Title.

PZ8.1.P2234 No 2001
398.2'0973--dc21

                                                    2001023221

Printed in Hong Kong

9 8 7 6 5 4 3 2 1

CONTENTS

North American Tales

North America is an enormous continent. It is two and a half times as big as Europe. There are icy, cold lands in the north and towering mountain ranges in the west. In the center of North America are wide, flat plains of wheat, and in the south are hot, dry deserts.

The first people to live in North America came from Asia about 15,000 years ago. Their descendants were the people we call Native Americans. They formed many different groups such as the Cheyenne, Pawnee, Iroquois, and Navaho. Each group had its own language and way of life. Some were farmers, others were hunters or warriors. Native Americans shared a great respect for all living things—from people, animals, and insects to plants, water, and the earth itself.

**Europeans began to settle in North America about
500 years ago. Spanish and Portuguese explorers came first,
and they called the Native Americans "Indians" because
they thought they had discovered India. Dutch, Swedish,
and English people also came to live in North America.**

These white people settled all over the country, fighting
the Native Americans and driving them off their land.
Today, there are not many Native Americans left.

The European settlers brought black people from
Africa to work as slaves on North American cotton
and tobacco plantations. Slavery was only abolished
in 1865. Many of the people living in the United
States today are descended from these Africans.

All these groups brought their own
myths and legends to North America.
You can read new versions of these
favorite tales in this book.

TALES OF THE GREAT HARE

Long, long ago, when all the world was water, Michabo the Great Hare went swimming. He dived down . . . down . . . down . . . deeper than he had ever swum before. The waters were cold and dark, but the Great Hare kept on swimming.

He swam so far and so deep that he reached the bottom of the ocean and felt the sandy sea bed beneath him. He picked up a single grain of sand and closed his paw tightly around it. The Great Hare then pushed off with his powerful legs and went swimming back upward, his long ears streaming out behind him.

After a long time, Michabo's head broke through the surface and he floated on the ocean in the sunlight. He opened his paw and watched the tiny grain of sand bob away from him on the waves. Before the Great Hare's eyes, the single grain became a thousand grains . . . then the thousand grains became a million grains . . . then the grains formed a patch of land that grew into an island, then a country, then a continent. And still the land kept growing and growing....

Michabo didn't know exactly how big the land became, but one day a wolf cub began to trot across it from one side to the other. The cub loped along for days and weeks and months and years . . . and by the time he had grown into an adult wolf, he still couldn't see the other side of the land. More years passed, and the wolf kept running, but when he was finally old and gray and could run no further, the end of the land was still nowhere in sight. *That's* how big the land was.

The land that the Great Hare had created was in fact what we now call Earth. Many peoples of different tribes and races came to live on it, but the Algonquian-speaking tribes who lived in northeastern America always thought of Michabo as their special friend.

Once, the Great Hare was resting on a riverbank when a boy came down to the river to fish. He threw his spear at the silvery, darting creatures time and time again—but they were always too quick for him.

Michabo was so busy watching the boy that he didn't feel a spider creep onto his head. By the time Michabo noticed, the spider had spun a web between his ears!

The web gave the Great Hare an idea for an easier way to catch fish. He took some string and wove a web just like the spider's—except much bigger.

Next he threw his web into the river. When he pulled it out, it was filled with wriggling fish! Because of Michabo's brilliant idea, the people of northeastern America were never short of fish again.

Another time, the Great Hare was lazily playing with a stick when two of his friends walked by. Michabo used his stick to doodle a picture of the man and woman in the earth. He watched the couple go into the forest —and he drew a picture of that, too. Then the Great Hare saw the man and woman coming out of the trees with handfuls of herbs—and he drew a third picture showing just that. When the man and woman walked back past Michabo, they saw the drawings and were delighted.

"Those pictures show us and what we did!" they laughed. "Anyone who came across them would be able to read them and know what they meant." And that's how the tribes of northeastern America came to use picture writing.

Michabo the Great Hare taught his human friends many other useful things. They were always sad in the winter when the Great Hare went away to his home in the east. There he slept for a long time, but every spring he returned to his people, and they were glad.

THE GIANT WOODCUTTER

When Paul Bunyan was born, he had enormous hands and feet and an even bigger appetite!

"That child's going to shoot up into a tall 'un!" said proud Mrs. Bunyan. That's exactly what Paul did—faster and taller than she ever imagined.

As a toddler, Paul was as big and strong as a full-grown man. At eight years old, Paul was so huge that he once sneezed and blew out all the windows in the house. By the time he was old enough to grow a beard everyone had lost track of how tall he was because there was no tape measure long enough to measure him!

When Paul Bunyan finally stopped growing, he was a giant. He was such an agreeable fellow, though, that no one was ever afraid of him. In fact, Paul decided to use his great height for the good of the whole town. Everyone needed wood: wood for buildings and furniture, wood for wagons and stables, wood for railroad tracks, wood for telegraph poles so people could send messages . . . the list was endless. Every day the townsmen struggled to cut down the enormous, heavy trees in the forest.

Paul Bunyan decided to lend a hand. He could cut down a tall pine tree with just three swishes of his ax!

The giant set up the finest logging camp for miles around. He chopped down trees so fast it took more than twenty barrels of ink every week to list them all.

When the townspeople had all the wood they could ever want, Paul Bunyan took orders from far and wide across the country. He had to employ hundreds of extra woodcutters to chop wood. There were so many men working at Paul Bunyan's logging camp that Paul had to fill a whole lake with pea soup to feed them. What's more, he had to use such a gigantic pan to fry their pancakes that it could only be greased if two cooks strapped hams to their feet and skated around inside it!

Perhaps the most unusual thing of all was Paul Bunyan's pet—an ox called Babe. Somehow Babe grew to be as enormous as Paul himself and, stranger still, he was bright blue! It was Babe who dragged the trees down the road to the river, so they could be floated downstream to be chopped up at the sawmill. There was just one problem—it was very hard to get the tall, straight trees down the twisting, zigzagging road.

It wasn't a problem for long.

Paul Bunyan tied Babe's harness to one end of the road and made the ox walk and walk and pull and pull until all the bends and corners were stretched right out. Soon the road was straight from start to finish, and the countryside was changed forever!

According to legend, this wasn't the only time Paul Bunyan helped shape North America. People say the giant once strode along dragging his pickax behind him, and the rut that it left in the ground was what we now call the Grand Canyon. It's over two hundred miles long and, in places, more than a mile deep! But maybe that's just a very tall story....

CROW AND THE PIECE OF DAYLIGHT

Near the very top of the world lie lands that are covered with snow and ice all year round. Howling winds blow up blinding snowstorms, and the air is bitterly cold. These frozen deserts are the lands that the Inuit people call home.

Once, the Inuit lands lay in darkness all year round. The Inuit people hunted polar bears in the darkness. They sat at ice-holes and fished in the darkness. They built igloos in the darkness. There was no dawn or dusk—the sky was always as black as midnight. In fact, the Inuit would never have known what the sun was if it hadn't been for their friend, Crow.

Crow traveled on his wings much, much farther than Inuit people could trek on their snowshoes. Crow flew to the very edges of the snow and ice, to where the blackness stopped. Beyond, the skies were bright with daylight. There, Crow soared in the warmth of the sun. He hovered over lands green with forests, purple with mountains, blue with rivers, and yellow with patches of sunshine.

15

At the end of the day Crow headed back into the dark lands of the Inuit and told the people about the brilliant lands of light.

"Daylight is amazing," Crow explained. "It's like lightning, but it isn't gone in a flash. It stays in the sky and lights the world with fabulous colors."

The Inuit men and women sat in their shadowy igloos and dreamed. They gazed into the blaze of their fires and stared at the light of their seal-oil lamps and imagined how wonderful it would be if their sky were bright all the time.

"Please bring us some of this daylight, Crow," they begged their friend.

No one saw Crow spring into the air and fly away, for his feathers were as dark as the blackness all around —but by next morning, he was gone.

As Crow flew, he thought about the Inuit people and how good and kind they were. There was so little food in their land that everyone always shared whatever they had. Crow knew that not all people were like that. He was worried that the people who lived in the world of daylight would not want to share any of their precious sunshine with the Inuit—not even a tiny piece.

Suddenly Crow burst out of the blackness and into the light. It wasn't long before he saw a village far below.

He swooped down to the largest, most important-looking house—the house of the chief. Crow landed on a windowsill and peered inside. His beady eyes twinkled when he saw the chief playing with his tiny grandson.

When the chief wasn't looking, Crow flew inside.

"Ask your grandpa for a piece of daylight to play with," he whispered to the little boy. Then he flew up to the roof and hid there.

"Grandpa! Grandpa!" the excited boy shouted. "Let me play with a piece of daylight!"

The chief thought that daylight was much too precious to be played with.

"I'll tell you a story instead," the chief smiled, but tears welled up in the little boy's big eyes. "I'll let you ride on my back," the chief promised, but the little boy began to sob. "I'll let you dress up in my big headdress with a thousand feathers," the chief said desperately, but the little boy wailed for the daylight.

Finally, the chief gave in. He snapped off a little piece of gleaming daylight. He tied a string to it so it wouldn't blow away and gave the string to his grandson, who beamed with delight. At that very moment Crow swooped down, grabbed the string, and flew away as fast as he could, trailing the glowing daylight behind him.

Far away, in the lands of snow and ice, the Inuit
saw a distant glimmer of light appear in the black sky.
Their hearts leaped for joy as they realized that Crow
was returning with a piece of daylight. Slowly, as Crow
flapped closer, the whole sky became lighter and brighter.
When Crow finally arrived, the snow and ice glittered
under bright skies as far as the eye could see.

From that day to this, Crow's small piece of daylight
has lit the Inuit lands for half of every year. The rest
of the time is as dark as the night. To this day the Inuit
people never harm crows—and now you know why.

WILEY AND THE HAIRY MAN

Wiley's mama knew all about things that were magic—
like the Hairy Man in the forest.

"The Hairy Man got your daddy, and if you're not careful,
Wiley, he'll get you too!" Wiley's mama often warned.

"I'll be careful," Wiley promised every time.

Wiley had never once so much as caught a sniff of
the Hairy Man. All the same, he felt better if he had his
two dogs with him when he went into the forest.

One day, Wiley was chopping wood when a pig ran
squealing by and his dogs raced after it. No sooner had
they disappeared among the trees than something huge
and hairy with sharp, pointy teeth came lumbering toward
Wiley. It was the Hairy Man! Wiley saw that the Hairy
Man had cows' hoofs and couldn't climb, so he shot up
a tree as fast as a squirrel.

The Hairy Man stood underneath the branches and
grinned a razor-sharp grin at Wiley.

"Come down and I'll show you some magic," he said.

"My mama's warned me all about you," yelled Wiley,
"and I'm not going to fall for that old trick!"

The Hairy Man stopped grinning. He picked up the ax that Wiley had dropped and began to hack away at the tree trunk. Soon the tree would come toppling down.

"Wait a minute, Mr. Hairy Man!" cried Wiley, thinking fast. "Let me say a short prayer before you eat me up."

"Very well," frowned the Hairy Man, leaning on the ax.

"Hoooo-eeeeee!" called Wiley at the top of his voice, and his two dogs came racing out from the trees, barking and snapping at the Hairy Man.

"Yikes!" gulped the Hairy Man, and he hurried away.

From that day on, the Hairy Man was determined to catch Wiley. He waited and waited until one day he saw Wiley in the forest without his two dogs. The Hairy Man sprang out of the bushes in front of Wiley. He waggled his hairy eyebrows at the boy and licked his hairy lips.

"Good afternoon, Mr. Hairy Man," said Wiley. "I'm glad I've bumped into you again. I've been thinking about what you said and I want to ask you something."

"Huh?" the Hairy Man shrugged, taken aback.

"You say you can do magic," continued Wiley. "So can you make things disappear—like all the rope in the neighborhood, for instance?"

"Of course," said the Hairy Man, scrunching up his eyes tightly, then opening them again. "There—it's done!"

"Oh good!" cried Wiley. "My dogs were tied up, but now they'll be free. Hoooo-eeeeee!"

"Yikes!" yelped the Hairy Man, fleeing into the forest.

Wiley's mama was very proud of her clever son—and she was excited too. She knew that if you could trick a monster three times, he'd have to leave you alone forever.

The next day the Hairy Man came to Wiley's house.

"Where's Wiley?" he bellowed, bashing down the door.

"If I give you my baby, do you promise to leave us alone forever?" demanded Wiley's mama.

"Ummm... yes," nodded the Hairy Man.

"Then take him," she said, pointing to Wiley's bed.

The Hairy Man's eyes lit up. He pulled back the sheets and grabbed—not Wiley, but a squirming piglet!

"This isn't your baby!" the Hairy Man growled.

"Oh yes it is!" laughed Wiley's mama. "I owned the piglet's mother and I own him too—only now he's yours!"

As the Hairy Man howled with fury, Wiley crawled out from his hiding place under the sofa.

"That's the third time we've tricked you, Hairy Man!" he grinned. "So now you have to leave us alone forever. Hoooo-eeeeee!"

"Yikes!" cried the Hairy Man, and Wiley's two dogs chased him all the way back to the forest.

Brer Rabbit and the Tar Baby

Brer Fox's greatest wish was to catch Brer Rabbit, but his traps and tricks always failed. That clever rabbit got the better of him every time.

But not this time! chuckled Brer Fox to himself. This time he had a *really* cunning plan. Brer Rabbit wouldn't be walking his bouncy walk and talking his cheeky talk and grinning his saucy grin for much longer!

That night, Brer Fox scooped and patted a big blob of sticky, black tar into the shape of a baby rabbit. Next he put the tar baby slap-bang in the middle of the road, where Brer Rabbit would see it on his morning walk to the lettuce patch. Finally Brer Fox hid in the ditch. All he had to do now was wait....

As the sun rose, Brer Rabbit came bouncing around the corner without a care in the world.

"Good morning!" Brer Rabbit greeted the tar baby. "It's a fine day, isn't it?"

The tar baby said nothing.

"I said, IT'S A FINE DAY, ISN'T IT?" hollered Brer Rabbit, just in case the tar baby was a little deaf.

The tar baby paid no attention.

Brer Rabbit felt rather angry. "Didn't your folks teach you any manners?" he asked.

The tar baby just ignored him.

This made Brer Rabbit hopping mad. "You've got until the count of three to say something—or else!" he yelled. "One . . . two . . . three!"

WHACK! Brer Rabbit smacked the tar baby—and his front paw stuck fast like glue. BLAM! He punched with his other paw—and that got stuck too! WHAM! Brer Rabbit kicked the tar baby—and found himself standing on one leg. THWACK! Brer Rabbit kicked again—and there he was, caught good and proper.

Brer Fox leaped out of the ditch and laughed and laughed.

"At last!" he cried with glee. "I'm going to eat tasty barbecued bunny tonight!"

Brer Rabbit let out a huge sigh of relief. "Oh, thank goodness!" he gasped.

Brer Fox's face fell. "Aren't you worried?" he growled.

"I can think of worse things than being roasted," smirked Brer Rabbit.

"All right, I'll drown you instead!" yelled Brer Fox.

"What a relief!" Brer Rabbit chuckled. "Drowning is still not all that bad."

"I'll string you up by the tail, then!" screamed Brer Fox.

"Do anything you like, Brer Fox," beamed Brer Rabbit, "just don't hurl me into the nasty, thorny briar patch. That would be the very worst thing you could do to me!"

"Right!" Brer Fox roared. "The briar patch it is, then!"

He grabbed Brer Rabbit by the ears, pulled him off the tar baby, and flung him into the air. Brer Rabbit came down—BUMP!—right in the thorny briar patch.

"Hee hee!" Brer Rabbit giggled, jumping up and dusting himself off. "I was born and bred in a briar patch, Brer Fox!"

Brer Fox had been tricked again! And while he howled with anger, Brer Rabbit bounced happily all the way to the lettuce patch.

THE SEARCH FOR HEALING

One winter, it wasn't just the snow and ice that came to stay at the village of the Iroquois people—a terrible sickness came too. The medicine man could do nothing to send it away. The sickness visited every wigwam and struck down several members in each family. Day after day, the sound of wailing filled the air as more men, women, and children breathed their last breath. Those still living grew exhausted and full of despair.

A warrior called Nekumonta watched each of his relations grow pale and then die from the terrible sickness. Before long, only his beautiful wife Shanewis was left— and she became ill too. Nekumonta did his best to care for Shanewis, but there was nothing he could do to make her better. The warrior felt helpless and angry.

Finally, the dreaded day arrived when Shanewis heard the voices of her dead ancestors calling her to take her place among them.

"My darling wife, you must not die and leave me," sobbed Nekumonta. "I will try to find the healing herbs of the Manitou—they will surely save you."

The Manitou was the greatest and most powerful of the Iroquois gods. There was an ancient story that the Manitou had planted herbs that would cure all ills—but no one knew where they grew.

"Beautiful Shanewis," Nekumonta whispered. "Hold on until I return from my journey!"

"I will try," Shanewis murmured weakly.

Nekumonta hurried off at once into the forest. He had no idea what the Manitou's healing plants might look like or where they might grow. Time after time, the desperate man plunged his hands into the freezing snow and dug deep, hoping to find the magic herbs.

The squirrels and deer and other creatures of the forest came out of their hiding places to see what Nekumonta was doing. They knew Nekumonta and weren't afraid of him. Though other men liked to hunt woodland animals for sport, Nekumonta had always treated all creatures with kindness and respect.

"Do you know where the Manitou has planted the herbs that will heal my people?" Nekumonta begged the creatures. None of them did. They crept back into the woods, sad that they were unable to help their friend.

Nekumonta searched for three days without success. By the end of the third day, he was freezing and starving.

He felt that he was no nearer finding the Manitou's herbs than before, and his tired head was full of thoughts of his beloved, dying Shanewis. As the cold sun sank down into the earth, he huddled under his blankets and sank into nightmares.

While Nekumonta tossed and turned, the animals of the forest held a meeting.

"This Iroquois warrior is a good human," said the bear.

"Yes," agreed the rabbit. "He has never once been cruel to us or damaged our homes."

"He deserves our help," the deer insisted. "What can we do?"

"We can't do anything," pointed out the beaver, "but the great Manitou himself can. Perhaps if we all ask him, he will realize how much every living thing wants Nekumonta to succeed in his brave quest."

Under the light of the full moon, the woodland creatures gathered in a clearing in the forest and cried out to the Manitou. The great god was extremely surprised to hear so many animals praying as one—particularly since they were asking for help for a human! The Manitou was moved by their loyalty and decided to grant their wish. He would help Nekumonta.

The Manitou sent the man a dream of his dying wife.

Softly, she sang a beautiful song that turned into a chorus of voices. The singing sounded like the music of a waterfall—and when Nekumonta awoke, he could still hear it.

"We are the healing waters...." the voices sang. "Find us . . . free us . . . then Shanewis and your people will be saved."

Suddenly, Nekumonta felt a surge of hope. He hunted high and low, and at last realized where the singing was coming from—an underground spring beneath his feet!

The forest creatures watched anxiously as Nekumonta scraped and scrabbled in the dirt, then—WHOOSH! A fountain of crystal water gushed into the air and bubbled away in a stream down the hillside. It splashed over Nekumonta, and he at once felt energy flooding into his body. The waters were magic!

Nekumonta's heart thumped inside his chest as he filled a skin bottle with the healing waters and raced back to his village. Soon, his lovely wife's eyes were bright and sparkling once again, and it wasn't long before every other sick person was healed too.

Nekumonta's kindness to the animals had been repaid, and everyone gave thanks with all their hearts to the great and good Manitou.

Tales of Davy Crockett

Davy Crockett wore a raccoon-fur hat and a deerskin jacket and lived all by himself in a log cabin in the forest. Most white settlers would have been terrified to live on their own in the wilderness, where they might meet Native American warriors or wild animals, but Davy Crockett seemed to be perfectly at home there.

People said that Davy Crockett could listen to a twig snap and tell you which animal was approaching. Stories spread that Davy Crockett could scare raccoons out of the trees just by grinning at them. News went around that Davy Crockett had killed a hundred and one bears in just one year, and that now bears turned tail and fled whenever they sniffed Davy Crockett's scent on the wind. But there was once a time when an animal got the better of Davy Crockett....

Davy Crockett was hunting in the woods one day when a raccoon scampered into the clearing right in front of him. The startled animal looked at the man and his gun and gulped.

"Eeek! It's Davy Crockett!" the raccoon yelped.

31

"Yep, that's right," replied Davy Crockett proudly, raising his gun and taking aim.

The raccoon took a deep breath.

"It would be an honor to be shot by you, Mr. Crockett," he announced. "You're the finest hunter these woods have ever known. Please fire away."

Davy Crockett was moved by the little animal's words. A tear glistened in his eye and he lowered his gun.

"Why, thank you very much," he told the raccoon. "After what you've just said, I don't think I'll shoot you after all."

"I'll be on my way then," called the raccoon cheerfully, "just in case you change your mind. Nice to meet you, Mr. Crockett!"

He scampered into the bushes and was gone.

Davy Crockett scratched his head and thought for a moment. He began to grin . . . then he threw back his head and laughed.

"I do believe that animal has just tricked me!" he guffawed. "That was the cleverest raccoon I've ever met!"

It was the one and only time that Davy Crockett was ever outsmarted. He was such an excellent hunter and tracker that he became famous far and wide. People started to call him the King of the Wild Frontier.

Davy Crockett became a hero to the people of North America. There's even a story of how he once saved everyone from being burned to a crisp when a shooting star called Halley's Comet came speeding toward the earth. People said that Davy Crockett grabbed the comet by its fiery tail and hurled it back into space, where it could cause no harm.

Davy Crockett died about two hundred years ago, but people today still enjoy telling tales of his amazing deeds—just like the ones here.

THE BABY AND THE GOD

The god Glooskap was a great warrior. He fought many wars and won every battle—not just with strength and bravery, but with wit and cunning too.

Glooskap spent many years away fighting. By the time he finally returned home, all his success and fame had made him rather big-headed.

"There's no one left in the world who doesn't fear me and won't obey me," he boasted.

"Oh yes there is," a woman piped up boldly. "I know someone who doesn't fear you—and he certainly won't obey you!"

"Who cannot have heard of the great Glooskap?" the annoyed god roared. "Who is it that dares not tremble before my name?"

"His name is Wasis," the woman calmly replied.

Hmmm, Wasis.... thought Glooskap. No, he was certain he had never heard of any warrior chief called Wasis on all his travels.

"Are you sure about this?" asked the puzzled Glooskap.

"Very sure," said the woman, with a twinkle in her eye.

"Wasis always does exactly what he pleases. He won't obey anyone—not even you, master."

"Then this Wasis must be very mighty," said Glooskap. "Is he as tall as the Kewawkqu'?" The Kewawkqu' were a race of giants. "Is his magic greater than the Medecolin?" The Medecolin were cunning magicians. "Is he as wicked as Pamola?" Pamola was an evil spirit of the night.

"Wasis is smaller than a goblin," said the woman. "He knows no magic, and there is no wickedness in him at all."

Glooskap was baffled. "So there's nothing special about this Wasis and yet he would challenge me!" he exclaimed. "I must see him for myself and teach him a lesson. Take me to him!"

"Wasis lives close by," said the woman. "Follow me."

The woman led the god among the wigwams belonging to her neighbors.

"Does Wasis live here in the village, among the ordinary folk?" asked the puzzled Glooskap.

The woman said nothing, but took Glooskap to her very own home.

"But this is where you live," exclaimed the god, surprised.

"Yes, and now Wasis lives here too," said the woman, and she led the god inside.

Glooskap peered around. "Where is he?" he shrugged.

"There," said the woman, and she pointed to a baby who sat on a rug, sucking a piece of maple sugar.

"That is Wasis?" cried Glooskap. He threw back his head and roared with laughter.

"Yes. Wasis is my son," smiled the woman. She knew that the great warrior god understood all about exploring and adventuring and fighting, but he had no idea about babies. He had never even met one before!

"Come here, Wasis," Glooskap cooed. He bent forward and held out his arms.

The baby's eyes opened wide, and he smiled at the strange man. Then he went straight back to sucking the piece of maple sugar he held in his fat little fist.

"Hmph," said Glooskap, somewhat annoyed. It was the first time anyone had ever failed to do as he said at once, but he didn't give up. The clever god cupped his hands to his mouth and whistled a beautiful bird song.

"Wonderful!" breathed the woman, but Wasis took no notice of Glooskap's music at all. It was as if he had suddenly gone deaf.

Glooskap was furious. "Come here!" he roared, waving his arms and stamping his foot. "Come here right now!"

Wasis looked up and stared straight at Glooskap.

For a moment, the god thought the baby was finally going to crawl toward him. Instead Wasis opened his mouth wide and began to scream.

"Be quiet!" raged Glooskap, as purple-faced as the baby. "Be silent and come here! I command you!"

The louder he shouted, the louder Wasis howled and wailed . . . and still he sat on the same spot on the rug.

Finally, Glooskap used his magic. He sang a song of enchantment so powerful that some people said evil spirits scurried to the depths of the earth to escape it. At last Wasis stopped howling and listened. He blinked his tearful eyes and began to grin. Soothed by Glooskap's music, the baby's little eyelids began to droop. Instead of crawling to the god, Wasis started to fall asleep.

With a roar of rage that could be heard for miles around, Glooskap gave up and stormed out of the wigwam. The woman smiled to herself and scooped Wasis up off the floor.

"I think you've taught the mighty Glooskap a lesson today," she whispered as she cuddled her baby close.

"Goo!" replied Wasis, trying to say Glooskap's name—and babies still say "Goo!" today, to remind us of the time when the greatest of gods was put firmly in his place by the smallest of children.

WHEN PEOPLE HAD WINGS

"I wonder how hot the sun is back home in Africa," whispered John. Sweat trickled off his forehead and soaked his shirt as he picked cotton off the plants.

Mary sighed. "There we'd be working for ourselves," she replied, under her breath. "We'd be feeding our families instead of making money for these white men."

Every day, it was the same. The African slaves were beaten awake at sunrise and marched to the cotton fields without any breakfast. By sundown, their whole bodies were aching and their stomachs burned with hunger. They hobbled back for a bowl of thin soup and a crust of stale bread, while inside the big house the master was served a delicious dinner from silver dishes.

"I wish I had my papa's wings so I could fly away," murmured Tom.

"Whatever do you mean?" whispered John in surprise.

"I mean what I said—I wish I had my papa's wings," hissed Tom.

The slaves fell silent as they heard the pounding of the overseer's horse coming up close behind them.

39

The overseer rode up and down the fields all day, checking that the men and women were working hard. Anyone found talking was silenced by a lash of his whip.

The overseer moved away, and Tom whispered, "Didn't you know that when our papas and mamas lived in Africa, they had wings and could fly?"

Mary and John shook their heads in amazement.

"Oh yes," sighed Tom. "They had beautiful, black wings, and they soared through the skies, free as birds."

"What happened to them?" whispered Mary.

"When the white people forced our mamas and papas onto boats and brought them across the ocean, they were so miserable that they lost the power to fly," explained Tom sadly. "Their wings shriveled up and died, and when we were born, we didn't have wings at all."

All day long, Mary and John thought about flying. They wondered what it would be like to hover in the heat of the African sun. They tried to imagine how it would feel to swoop through the bright skies of their homeland.

That night, when John was asleep in his hard bunk, he dreamed that he cried out loud for wings. A wrinkled, old man appeared by his bedside and shook him awake.

"I am the One Who Remembers," the man said, his eyes shining. "Wake the others and come outside."

Soon all the slaves were gathered in the chill moonlight.

"Join hands," instructed the old man, so the men and women huddled closer and linked their fingers. The old man shut his eyes and chanted some mysterious words.

A shiver ran down the spine of each and every slave as their back muscles seemed to ripple with strange energy. Suddenly they felt huge wings rip through the cloth of their shirts. With a few powerful wingbeats, all the slaves soared into the dark air. They were free!

Next morning, the overseer found that his slaves had vanished without a trace. Only a handful of long black feathers were tumbling in the wind over the dust....

THE STORY OF DEATH

The Maidu people believe that at the beginning of time, the gods Kodoyanpe and Coyote floated on the surface of a huge ocean and created all things. They created the land, plants, and animals. They created people—so many that the two gods became worried.

"What shall we do?" Kodoyanpe said to Coyote. "Our people are having lots of children. They bring joy and happiness and hope for the future, but there isn't enough space on the earth for so many people."

"Plants die," Coyote said thoughtfully. "Animals die. Why shouldn't humans die too?"

Kodoyanpe looked worried. "That would be cruel," he said in a low voice. "It would make humans very unhappy if their loved ones went away from the earth forever."

"Well, have you got a better idea?" snapped Coyote.

The Caddo people believe that Kodoyanpe decided that people should die after all, but that their spirits should leave the earth and stay in a special house. After a while the spirits would be able to come back to life in a new body.

43

However, the Caddo people say that when the first person died, Coyote turned himself into a wild dog. He blocked the entrance to the house and wouldn't let the spirit inside. The soul wandered the skies until he came to the land of the spirits, from which there is no return. This is the way it has been with humans ever since.

The Maidu people tell a different story. They say that the quarrelsome Coyote just argued and argued with Kodoyanpe until he got his way. People would die and leave the earth forever.

We will never know which story is true, but it doesn't really matter. The result was the same—death came to mean the end of life on Earth for humans.

After Kodoyanpe and Coyote had argued over death, they went their separate ways to live among humans. Kodoyanpe loved living among the people he had created. Coyote, on the other hand, didn't care about anyone— until his son was born. As Coyote looked at his little boy, he felt love for the first time. As the child grew older, he brought Coyote joy and happiness and hope for the future, just as Kodoyanpe had said children did.

But one day Coyote's son was bitten by a snake.

"Father, I am dying," the little boy gasped as the poison chilled his veins. "Help me!"

Coyote threw back his head and howled in anguish. He snatched up his son and raced like the wind over the earth until he found Kodoyanpe.

"I was wrong about death!" Coyote wailed. "Death shouldn't be the end of things. I can't bear to live without my son. Is there anything we can do?"

Kodoyanpe wept as he watched life leave the little child's body.

"It was you who wanted death to be the end," he whispered. "What's done is done. We cannot undo it."

Coyote howled another bloodcurdling cry of pain. As he did so, he crouched down on all fours. His body became hunched and hairy. He scowled an ugly sneer, and his tongue lolled out of sharp-toothed jaws. Claws sprang from his feet, and his eyes flashed with yellow fire.

"I think you knew how to save my son," Coyote snarled. "You just didn't want to! I won't forget this, brother!"

Coyote sprang across the world in his new form, filled with bitterness and hatred. He roamed the earth, stirring up trouble and making as much mischief as possible.

Kodoyanpe shook his head in despair when he saw the misery Coyote was causing his people.

"Build a giant canoe," Kodoyanpe secretly told the people. "Make it big enough for everyone to fit inside."

45

All over the earth, people worked together to build the enormous canoe. They clambered aboard, and then Kodoyanpe created the biggest storm the world had ever seen. The sky was blotted out with black storm clouds. Lightning zigzagged through the darkness, and thunder crashed through the heavens. Waterfalls came pouring out of the skies, swelling the rivers and lakes and seas until they joined together as one huge ocean. Kodoyanpe was going to drown Coyote by flooding the world.

When the rains finally stopped, Kodoyanpe looked all around and saw that everywhere was water. The only land that remained poking up above the waves was the very tips of the highest mountains—and there on one of them was the laughing Coyote! The cunning god had disguised himself as a person and crept into the canoe with everybody else. As the giant boat had swept past a rocky peak, crafty Coyote had jumped out onto it.

Kodoyanpe's plan had failed. To this day Coyote is still out there somewhere—which is why we have evil in the world.

INDEX